U.S. DEPARTMENT OF THE TREASURY

PROGRESS REPORT

Study of

I0503124

MARKING, RENDERING INERT AND LICENSING OF EXPLOSIVES MATERIALS

- 1997 -

<u>Department of the Treasury Press Report</u>

PROGRESS REPORT
STUDY OF MARKING, RENDERING INERT,
AND LICENSING OF EXPLOSIVE MATERIALS
Table of Contents

EXECUTIVE SUMMARY

The Antiterrorism and Effective Death Penalty Act of 1996 (the Act) was enacted on April 24, 1996. Section 732 of the Act mandates the Secretary of the Treasury (the Secretary) to conduct a study of: the tagging of explosive materials, rendering common chemicals used to manufacture explosive materials inert, imposing controls on certain precursor chemicals and, State licensing requirements for the purchase and use of commercial high explosives. On September 30, 1996, an amendment included in the Omnibus Consolidated Appropriations Act for Fiscal Year 1997 (Amendment) provided the necessary funding for the study and added a requirement that the Secretary, in consultation with the Attorney General, concurrently report to Congress on the possible use and exploitation of new prevention technologies.

The Amendment also calls for the Secretary to enter into a contract with the National Academy of Sciences (NAS) to conduct a separate study of tagging smokeless and black powders. That separate NAS study on black and smokeless powders is not part of the general taggant study and is not addressed in this progress report.

The Secretary has delegated the responsibility for conducting the study (excluding smokeless and black powders) to the Bureau of Alcohol, Tobacco and Firearms (ATF), the Federal agency responsible for enforcing the Federal statutes regarding the criminal use of explosives and the regulatory controls placed on the manufacture, importation, sale, and distribution of explosive materials in the United States. On April 26, 1996, ATF established its Explosives Study Group ("Study Group") consisting of nine members with a total of 193 years of experience in law enforcement, explosives regulatory enforcement, and forensic chemistry.

In initiating its study (the Study), ATF reviewed the magnitude of the U.S. explosives industry; the events leading up to the pilot taggant program conducted between 1977-79; the Office of Technology Assessment analysis of that pilot program; and the existing Swiss program for tagging explosives. In addition, ATF forged a partnership with the explosives, chemical, fireworks, and fertilizer industries to ensure broad public input and support for solutions to an immensely complex problem.

ATF capitalized on its 26 years of experience regulating the explosives industry and investigating bombings. In addition, it relied on the knowledge gained fromthe ATF-hosted 1995 International Explosives Symposium which addressed issues concerning the identification and detection of explosives, neutralizing common chemicals used in the manufacture of improvised explosive devices, and developing new and innovative ways to combat the criminal use of explosives. As a result of the Symposium, and in an effort to demonstrate an example of possible voluntary controls, ATF teamed with The Fertilizer Institute to design a program to alert the fertilizer industry to the issues of security, knowledge of purchasers, and the need to recognize thefts. The program, "Be Aware For America," has been a source of invaluable information on voluntary efforts.

To supplement the Study, ATF commissioned NAS and The International Fertilizer Development Center to study specific issues as mandated by the Act. ATF also conferred with all other Federal agencies with a stake in the tagging issue, as well as several foreign governments who share our same concerns for answers to the taggant questions.

As the Study continues, ATF is assembling experts in various fields to evaluate a host of proposals related to the issues of tagging explosives and controlling or rendering inert common or precursor chemicals. National research facilities which can conduct a full range of tasks related to those issues are also being identified.

ATF intends to issue another report in 1998. To this end, ATF will commission research in a variety of areas to include: an analysis of the processing differences between U.S. and Swiss explosives manufacturing; the potential environmental impacts of adding taggants to certain explosives; costs of a tagging program in the U.S.; assessing new technologies; and assessing whether vapor detection marking agents currently included in plastic explosives can be used in other selected explosives. Finally, ATF will work with the Drug Enforcement Administration to analyze their drug-related chemical watch program for possible utility in the explosives arena and the chemical industry to develop other enhanced controls. ATF will also actively participate within the Department of State's (DOS) Technical Support Working Group (TSWG) to evaluate new and existing detection technologies, and evaluate options that are a result of the ongoing TSWG study regarding ammonium nitrate (AN) desensitization alternatives. Because this work is ongoing, the following is an interim status report on the Study.

At this stage of the Study it is clear that while the tagging of explosive materials holds great promise for the prevention and investigation of certain incidents, there are remaining complexities surrounding the issue. Any effort which is to have a measurable impact on the prevention and investigation of bombing incidents must be an integrated one, involving the effective regulation of explosives and explosive materials, the effective enforcement of those regulations, and the effective application of cutting edge technologies.

I. SCOPE OF STUDY

In response to recent explosive-related incidents, the Antiterrorism and Effective Death Penalty Act of 1996 (the Act) was enacted on April 24, 1996. Section 732 of the Act mandates that the Secretary of the Treasury (the Secretary) conduct a study of: the tagging of explosive materials for purposes of detection and identification; the feasibility and practicability of rendering common chemicals used to manufacture explosive materials inert; the feasibility and practicability of imposing controls on certain precursor chemicals used to manufacture explosive materials; and, State licensing requirements for the purchase and use of commercial high explosives.

On September 30, 1996, the Omnibus Consolidated Appropriations Act for Fiscal Year 1997 provided the necessary funding for the study and added a requirement that the Secretary, in consultation with the Attorney General, concurrently report to Congress on the possible use and exploitation of new prevention technologies, including vapor detection devices; computed tomography; nuclear quadropole resonance; thermal neutron analysis; pulsed fast neutron analysis, and other technologies upon which recommendations to the Congress may be made for further study, funding, and use of the same in preventing and solving acts of terrorism involving explosive devices. [The amendment also calls for the Secretary to enter into a contract with the National Academy of Sciences (NAS) to conduct a study of the tagging of smokeless and black powders by any viable technology for purposes of detection and identification. That separate NAS study on black and smokeless powders is not part of the general taggant study and is not addressed in this progress report.] (See Appendix A-1, the Act.)

The Secretary has delegated the responsibility for conducting this study to the Bureau of Alcohol, Tobacco and Firearms (ATF), the Federal agency responsible for the criminal and regulatory enforcement of the Organized Crime Control Act of 1970, Title 18 U.S.C., Chapter 40, and its implementing regulations contained in 27 CFR Part 55, Commerce in Explosives. The Federal explosives regulations encompass the licensing, storage, and recordkeeping requirements for explosive manufacturers, importers, dealers, and users.

ATF has made significant progress in this study. On April 26, 1996, ATF established its Explosives Study Group ("Study Group") to examine the issues surrounding the use of identification and detection taggants in explosives. The Study Group consists of nine members, with a total of 193 years of experience in law enforcement, explosives regulatory enforcement, and forensic chemistry. The Study Group first researched the work that had already been done in the area of taggants by ATF and other entities. In addition, the Study Group has received areport from the International Fertilizer Development Center (IFDC) addressing issues relating to fertilizers. ATF contracted with the National Academy of Sciences (NAS) for an independent and parallel study of the same issues it was delegated to examine. (This NAS study is totally separate from the study referred to in footnote 1, on page 1.) The results of that report have not been included. Any findings, conclusions, or recommendations from NAS will be considered in formulating the additional work to be performed in the second phase of this study. ATF consulted with foreign governments and manufacturers regarding their experiences with tagging explosives or rendering precursor chemicals inert. ATF has also consulted extensively with representatives of the explosives, fertilizer, and chemical industries to take advantage of their technical expertise on this issue, and to receive their opinion on the economic impact of requiring taggants in explosive materials, the feasibility and practicability of rendering common chemicals inert, and placing controls on certain precursor chemicals used to manufacture explosive materials.

This progress report sets forth the steps taken by the Study Group to date to gather information on these complex issues. Since further work remains to be done, this progress report does not contain any final results, nor does it make any recommendations for legislation. It does, however, summarize the extensive work done by ATF, focus the scope of the remaining study, and identify 15 specific actions left to be taken to achieve the objectives of the study. It also sets out the time frames for completion of the study.

Another report providing the most recent findings of the Study and potential recommendations for legislation will be issued in 1998. The final report mandated by the Act, including the results of the Study and any recommendations for legislation will be issued 30 days after completion of the Study. In addition, if the requisite findings are made in accordance with the Act, the Secretary may also propose regulations to require the addition of tracer elements to explosive materials.

II. SUMMARY OF WORK COMPLETED

A. BACKGROUND

Since the 1970s, there have been various proposals to mark or "tag" explosive materials for purposes of pre-blast detection or post-blast identification for use in investigating criminal bombings or attempted bombings. A "taggant" or "tracer element" can be a solid, liquid, or vapor emitting substance put into an explosive material for the purposes of detection or identification before an explosion occurs or for identification after an explosion occurs.

A detection taggant is designed to be identified by a suitable detection device even when the explosive is contained or concealed in a package. Detection machines at airports or other sites could signal any effort to introduce tagged explosive materials into the area.

Since 1971, the Federal explosives law and regulations require licensed manufacturers of explosive materials to legibly identify, by marking, each cartridge, bag, or other immediate container of explosive materials manufactured for sale or distribution. The marks required must identify the manufacturer and the location, date, and shift of manufacture. Licensed manufacturers must keep records of sale or distribution by the marks of identification (date-shift-code), description (brand name), size, and quantity. Licensed distributors also must maintain records of acquisition and disposition of explosive materials. A trace can be undertaken to compile a list of the last legitimate purchasers of all or part of a particular batch of explosives manufactured on a specific date and during a specific shift. However, while this has proven useful in investigations of criminal bombings, its utility is limited to instances where the explosive is recovered before detonation, or, in some cases, where a low-order detonation does not destroy the cartridge.

A need therefore remains for an appropriate taggant to be placed in the explosive material itself, designed to detect an explosive material prior to detonation and/or to survive a blast and, through a code with the same identification information that is required on the cartridge, bag, or other immediate container, enable traces to be made when that packaging is destroyed in the blast.

A-1. Advisory Committee on Explosives Tagging - 1973

In 1973, the Advisory Committee on Explosives Tagging was formed. The Committee, chaired by ATF, was composed of representatives from Federal agencies and explosives industry representatives with an interest in, or involved in,

conducting research into the viability of tagging explosives for detection or identification.

As the Chair of the Advisory Committee, ATF coordinated the efforts of the interested parties and selected a contractor to conduct a pilot research program. In 1976, ATF selected the Aerospace Corporation as the system technical manager for the identification and detection tagging pilot program. Aerospace was selected primarily due to its work on similar projects for the Law Enforcement Assistance Administration, Department of Justice (DOJ), and the Bureau of Mines (Department of the Interior).

The Minnesota Mining and Manufacturing Company (3M) developed an identification taggant for Aerospace to test. This taggant is a color-coded, polymer "microchip" consisting of ten layers, including a magnetic layer and a fluorescent layer. Four manufacturers of explosives--Atlas Powder Company, E.I. duPont deNemours and Company, Hercules, Inc., and Independent Explosives Company, Inc.,--participated in the pilot program by adding the 3M taggant to the manufacture of approximately 7.5 million pounds of dynamite, slurries, water gels, and emulsions. These companies ultimately withdrew from the program, citing their concerns about the safety of taggants due to compatibility testing which allegedly indicated that the taggants might destabilize certain explosive products. The companies were further concerned about an explosion that occurred in July of 1979 at the Goex manufacturing plant in East Camden, Arkansas. While Goex asserted that taggants in boosters which were being reworked caused the explosion, it has never been established that taggants were involved in, or the cause of, the explosion.

The pilot program resulted in one concrete example of the utility of identification taggants. In 1979, a truck bombing in Baltimore, Maryland, resulted in the death of the driver. Water gel used in the bombing was manufactured during the pilot program and contained 3M identification taggants. In the ensuing investigation, ATF developed a suspect, traced the water gel using the 3M taggant, and found that the suspect had purchased the same type of water gel used in the bombing from an explosives dealer in Charleston, West Virginia. The successful trace of the water gel using the 3M taggant provided important evidence to support the Federal charges against the defendant. The defendant was convicted and sentenced to 30 years' incarceration. The U.S. Attorney responsible for the case formally stated in a letter to ATF that "the investigation and prosecution succeeded largely because of the discovery of taggants at the crime scene."

A-2. Office of Technology Assessment Report - 1980

In 1979, the Senate Committee on Governmental Affairs asked the Office of Technology Assessment to review the data on taggants in conjunction with proposed legislation that would have mandated the use of identification and detection taggants in explosive materials. The OTA was asked to review the available data and address the safety of taggants in the production, storage, and handling of explosive materials; the effectiveness of the tagging program in deterring crime and aiding in criminal investigation and prosecution; the regulatory impact of requiring the use of explosives taggants (including the cost to the industry and consumer); the potential effects of a partial application of tagging requirements; issues relating to the survivability of taggants; and possible alternatives to tagging explosives and initiators.

The study issued by OTA in 1980 concluded that there were unresolved issues regarding the compatibility of taggants with different types of explosive materials. The OTA study found that the testing done to date created a reasonable presumption that the 3M identification taggant was compatible with dynamites, water gels, slurries, emulsions, and black powder. It also stated, however, that there was evidence of increased reactivity, and thus a presumption of incompatibility, with at least one form of smokeless powder and at least one cast booster composition. OTA concluded that it was not yet possible to arrive at presumptions about the compatibility of the 3M taggant with blasting caps or detonating cord, or about the compatibility of detection taggants with any commercial explosive. OTA further stated that, even for products such as dynamites where no evidence of incompatibility existed, further testing was required before it could definitely be concluded that taggants were compatible with, and could safely be added to, all such explosives.

Assuming that stability questions were successfully resolved and that technical development was successfully completed, the OTA determined that "both identification taggants and detection taggants would be useful law enforcement tools against most terrorist and other criminal bombers." At that time, the Report concluded that "identification taggants would provide a quantum increase in utility for those bombings significant enough to warrant a thorough investigation, while detection taggants would provide that increased utility in protecting those potential targets sufficiently important to warrant a detection taggant sensor." However, the report also noted that countermeasures existed that could enable bombers to evade the effects of a tagging program. Because available countermeasures require varying degrees of specialized knowledge, and some involve significant risks, most bombers would probably not avail themselves of those countermeasures. In summary, the OTA concluded that a taggant program "would probably retain substantial law enforcement utility," however, its utility against the most sophisticated of terrorists and professional criminals was considered open to question.

The OTA noted that "the cost of a taggant program can vary by almost an order of magnitude, depending on the implementation plan." Variables in such a plan could include how often the manufacturer would be required to change its unique taggant, and how many detection sensors would be employed. Another factor was whether all explosive materials would be tagged, or whether ANFO would be excluded. The OTA identified a "baseline" plan that would increase the cost of explosives to the ultimate user by approximately 10 percent.

The overall conclusion of the OTA report was that, "Additional information is required on all aspects of the analysis - technical efficacy, safety, cost, and utility."

A-3. Appropriations Restrictions and Developments Since 1980

In October 1980, the House Treasury, Postal Service, and General Government Appropriations Subcommittee voted to prohibit ATF from expending additional funds on tagging research, development, or implementation. Intermittently, between fiscal years 1980 and 1993, ATF appropriations contained language stating that "none of the funds appropriated herein shall be available for explosive identification or detection tagging research, development, or implementation." Thus, prior to enactment of the Act, ATF had not expended funds on tagging research, development, or implementation since 1979.

In 1995, under its general authority to regulate explosives, ATF participated in two significant programs that produced greater awareness of explosives safety and law enforcement issues. A brief explanation of these events follows.

A-3(a). International Explosives Symposium

In September 1995, ATF hosted an International Explosives Symposium in Fairfax, Virginia, which assembled a diverse group of over 100 participants and 35 speakers from North America and Europe. The objectives of this symposium were to address issues concerning the identification and detection of explosives, the neutralization of common materials used in the manufacture of improvised explosive devices, and the need to develop new and innovative ways to combat the criminal use of explosives.

Symposium participants included individuals from the scientific and technical communities, the public and private sectors, forensic experts, industry representatives, and law enforcement officials. They were provided the opportunity to exchange information concerning existing and theoretical technologies and develop a deeper understanding of the complexities of explosiveuses and controls. The papers presented by the various speakers covered such topics as:

"Ammonium Nitrate and the Provisional Irish Republican Army"

"The ATF Canine Explosives Detection Program"

"Counter-Terrorism Explosives Research"

"Desensitizing Ammonium Nitrate Fuel Oil"

"Detection Tagging of Packaged Cap Sensitive Explosives"

"Research on Nitrate Fertilizers and the International Fertilizer Development Center Capabilities"

"The National Pilot Test Program for the Identification Tagging of Explosives"

"Reduced Terrorism--The Use/Effectiveness of Ammonium Nitrate Explosives"

The papers were compiled and distributed throughout the Government and scientific communities, and to international participants.

A-3(b). "Be Aware for America" Program

Following the 1995 Symposium, The Fertilizer Institute (TFI) solicited ATF's assistance in developing a national awareness program entitled "Be Aware for America." The campaign was designed to help the entire fertilizer industry be alert to suspicious purchasers, heighten security, increase vigilance over storage and distribution, and ensure that persons are able to recognize the theft from, or the misreporting of, fertilizer product shipments. As part of this awareness program, ATF provides a toll-free hot-line number (1-800-800-3855) for reporting suspicious activities.

A-4. Detection Agents in Plastic Explosives

In 1991, the United States and 43 other participating nations signed the Convention on the Marking of Plastic Explosives for the Purpose of Detection, Done at Montreal on 1 March 1991. The Convention required all signatory nations to take necessary and effective measures to exercise controls over unmarkedplastic explosives. The Convention represents the response of the international community to the threat posed to the safety and security of international civil aviation by virtually undetectable plastic explosives in the hands of terrorists. Such explosives were used in two airline tragedies: the bombings of Pan Am Flight 103 over Lockerbie, Scotland, in December 1988 and UTA Flight 772 in September 1989.

Title VI of the 1996 Antiterrorism Act amended the Federal explosives laws to carry out the responsibility of the United States under the Convention. As of April 24, 1997, it is unlawful to manufacture, import, export, or possess any plastic explosive that does not contain a detection agent. Plastic explosives that were imported into or manufactured in the United States prior to that date may be lawfully possessed until April 24, 1999. There are further exemptions for the use by Federal law enforcement or military agencies, or the National Guard of any State, of plastic explosives manufactured or imported prior to that date. ATF published regulations implementing the new statutory requirements on February 25, 1997.

The detection agents required by law to be placed in plastic explosives are designed to render plastic explosives detectable by vapor detection equipment or specially trained canines. The law specifies four potential chemical marking agents. One of them (2,3 dimethyl--2,3 dinitrobutane - DMNB) was selected by the U.S. Army for use in marking U.S. military plastic explosives following satisfactory completion of extensive health, safety, environmental, and compatibility testing. DMNB is commercially available although its price is currently very high. The four chemical marking agents are not identification agents: they do not identify the manufacturer or importer of the plastic explosive.

B. OVERVIEW OF THE EXPLOSIVES INDUSTRY AND THE MISUSE OF EXPLOSIVES IN THE UNITED STATES

The Study Group first reviewed available data concerning the American explosives industry, as well as information regarding the unlawful use of explosives. [The Study Group analyzed available statistics on explosive-related incidents, in order to establish an effective reference point from which to assess and address the criminal use of explosives. ATF 's Explosives Incidents System (EXIS), developed in 1975, is a computerized repository for statistical data on explosives incidents. Data compiled in EXIS is primarily reported by ATF, the Federal Bureau of Investigation (FBI), and the U.S. Postal Service, and is published in ATF 's Explosives Incidents Report (EIR) annually. The information contained therein is not exhaustive of all incidents which occurred during a given year, but is reflective of those incidents reported. However, the data is considered highly representative and sufficient to permit valid chronological, geographical, and trend analyses. The Study Group analyzed statistics contained in the EIR covering the period 1976 through 1995. Statistics for 1996 are not yet available.]

ATF recognizes that the legitimate uses of explosives are many, and that the explosives industry is of significant economic importance to this country.

B-1. U.S. Production of Explosive Materials

The U.S. Department of the Interior, U.S. Geological Survey (USGS), 1995 Annual Review, reports that the total production of high explosives and blasting agents in the U.S. in 1995, which includes commercial explosives imported for industrial uses, was 2.28 million metric tons (approximately 5 billion pounds). According to the USGS, the "principal distinction between high explosives and blasting agents is their sensitivity to initiation. High explosives are cap-sensitive, whereas blasting agents are not." Table S.1 reflects 1995 [Statistics for 1995 are the most recent available.] production in pounds and production classifications referenced by the USGS and used by the Institute of Makers of Explosives (IME). The terms used in the table are defined as follows:

High Explosives are defined as explosives which are characterized by a very high rate of reaction, high pressure development, and the presence of a detonation wave in the explosive.

Blasting Agents are defined as any material or mixture consisting of fuel and oxidizer intended for blasting, not otherwise defined as an explosive, provided that the finished product, as mixed for use or shipment, cannot be detonated by means of a No. 8 test blasting cap when unconfined.

Ammonium Nitrate (AN) is classified as an oxidizer. An oxidizer is a substance that readily yields oxygen or other oxidizing substances to promote the combustion of organic matter or other fuel. AN alone is not an explosive material. However, Federal explosives storage regulations require the separation of explosive magazines from nearby stores of AN by certain minimum distances.

Bulk Mix is defined as a mass of explosive material prepared for use in bulk form without packaging.

Slurries are defined as explosive materials containing substantial portions of a liquid, oxidizer, and fuel, plus a thickener.

Water Gels are defined as an explosive material containing substantial portions of water, oxidizers, and fuel, plus a cross-linking agent which may be a high explosive or blasting agent.

Emulsions are explosive materials containing substantial amounts of oxidizers dissolved in water droplets surrounded by an immiscible (incapable of blending or mixing) fuel.

Table S.1	
Estimated U.S. Production of Commercial Explosives - 1995	
Product	**Average Pounds Per Year (000)**
High Explosives	
Permissibles - Grades of explosives approved by brand name, by the Mine Safety and Health Administration, as established by U.S. Bureau of Mines testing. (Primarily used in underground mining.)	7,524
Other High Explosives: All high explosives except permissibles.	80,080
Blasting Agents: All mixtures, regardless of density.	
Ammonium Nitrate Fuel Oil (ANFO)	965,800
Bulk Slurries, Water Gels, and **Emulsions**: All bulk slurries, water gels, emulsions and ANFO mixtures containing slurries, water gels and emulsions.	783,200
Unprocessed Ammonium Nitrate: Includes prilled, grained, and water solution (liquor) ammonium nitrate sold for use in the manufacture of commercial explosives.	3,190,000

Source: U.S. Department of the Interior, U.S. Geological Survey, 1995 Annual Review (See Appendix B-1).

According to the 1995 Annual Review, some explosives sales may be concealed under "unprocessed ammonium nitrate" to avoid disclosure of individual company proprietary data.

The IME and the International Society of Explosives Engineers (ISEE) report that 400 to 500 different explosive products are manufactured in the U.S. each year.

There are approximately 10,600 Federal explosives licensees and permittees in the U.S. A licensee is any importer, manufacturer, or dealer licensed under the Federal explosives laws. A permittee is any person who has obtained a Federal User Permit to acquire, ship, or transport explosive materials in interstate or foreign commerce. Of the total 10,600, 5,429 are licensees comprised of 313 importers, 3,365 dealers, and 1,751 manufacturers. The 1,751 manufacturers are comprised of 1,097 manufacturers of high explosives, 186 manufacturers of lowexplosives, 162 manufacturers of blasting agents, and 306 manufacturers of theatrical flash powders. The remaining 5,171 of the total 10,600 are permittees.

B-2. Analysis of Available Statistical Data Concerning Explosives Incidents

Of the approximately 4,011 thefts reported between 1978 and 1995, approximately 52 percent occurred at the user level. In addition to licensees and permittees, "users" are defined in the EIR as any individuals who purchase and use explosives within their State and are therefore not required to obtain a Federal license or permit. Explosives are generally not sold in less than case lots, which may result in purchases of more than is required by the purchaser.

Figure S.1

Percentage of Explosive Thefts as Reported by Licensees, Permittees, and Users

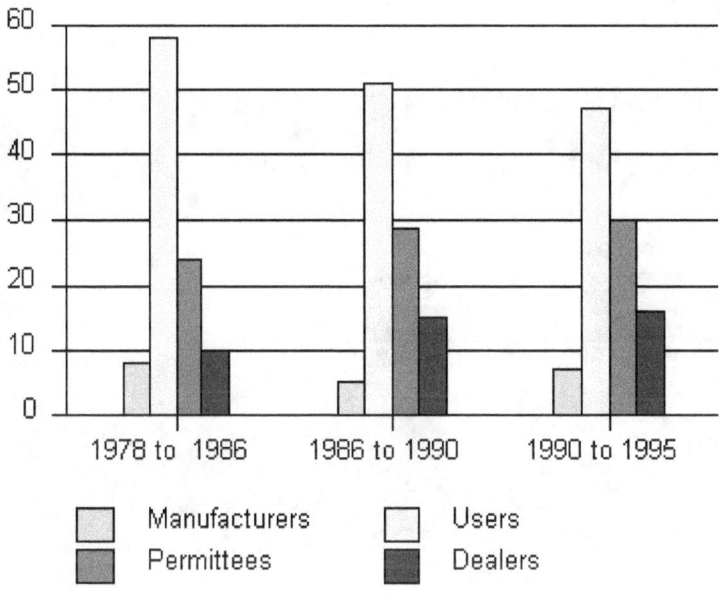

Source: Department of the Treasury, Bureau of Alcohol, Tobacco and Firearms' Explosives Incidents Reports.

From the 5-year period 1976-1980, to the 5-year period 1991-1995, the number of incidents of stolen high explosives and blasting agents decreased from 1,700 to 481. From the 5-year period 1976-1980, to the 5-year period 1991-1995, the number of pounds of stolen high explosives and blasting agents decreased from

636,238 pounds to 58,040 pounds. These decreases are primarily due to the explosives industry's improvements in security and compliance with ATF-administered security regulations which are enforced by regulatory inspections.

Figure S.2

Thefts in Total Poundage in 5-Year Intervals, 1976-1995

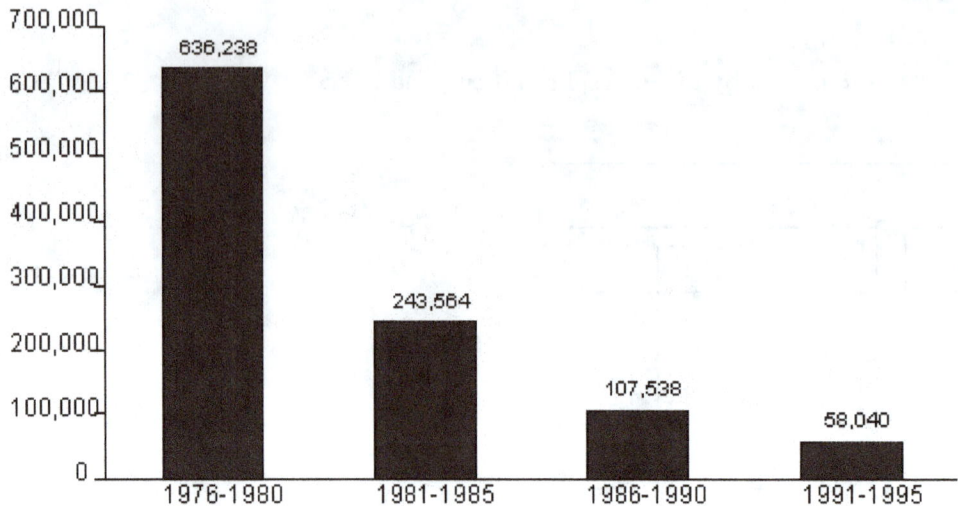

Figure S.3

Thefts by Incidents in 5-Year Intervals

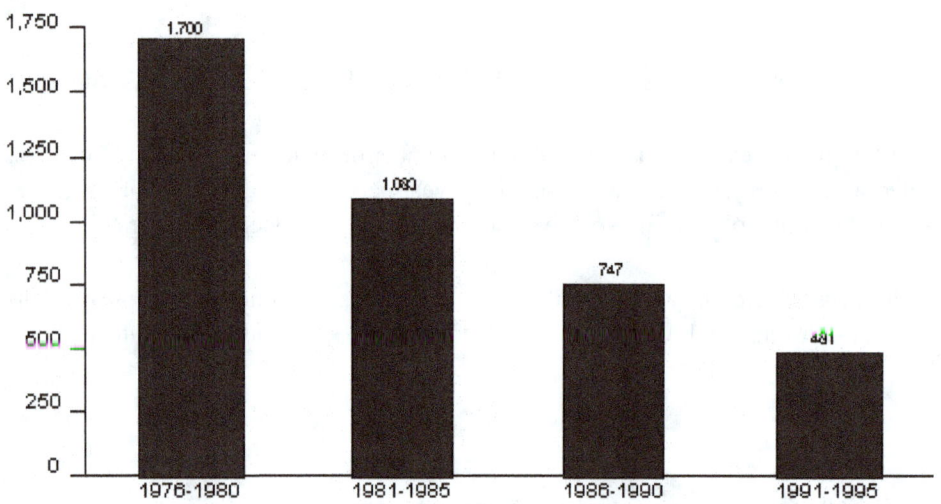

Source: Department of the Treasury, Bureau of Alcohol, Tobacco and Firearms' Explosives Incidents Reports 1985 and 1990, and Arson and Explosives Incidents Report 1994, 1995.

Statistics on the types of fillers used in attempted bombings and bombing incidents are available beginning in 1976 and for each year through 1995. The more recent and narrower grouping of figures, for the period 1991-1995, is frequently referenced by explosives industry representatives. The data reported for both time periods is represented in Table S.2 and in Figures S.2a, and S.2b.

Table S.2				
Fillers Used in Bombings and Attempted Bombing Incidents				
	1976 through 1995		1991 through 1995	
Filler	Incidents	Percentage	Incidents	Percentage

Undetermined*	8,705	24.6	2,588	18.2
Black or Smokeless Powders	8,463	23.9	3,175	22.3
Flammable Liquids	8,075	22.9	3,399	24.0
Photo Flash/Fireworks Powders	3,115	8.8	1,715	12.1
Chemicals	2,967	8.4	2,554	18.0
Other**	1,994	5.6	586	4.1
Dynamites and Water Gels	1,791	5.1	140	1.0
Blasting Agents	161	0.5	27	0.2
C4/TNT	55	0.2	15	0.1
Total	**35,326**	**100.0**	**14,199**	**100.0**

*The Undetermined category captures incidents in which fillers could not be identified through laboratory analysis, or incomplete data was reported.

**The Other category includes match heads, military explosives (excluding C4/TNT), improvised mixtures, flares, boosters, detonating cord, gases, blasting caps, PETN, RDX, HMTD, model rocket propellant, and smoke grenades.

Source: The Department of the Treasury, Bureau of Alcohol, Tobacco and Firearms, Explosives Incidents

Reports 1985 and 1990, and Arson and Explosive Incidents Reports 1994 and 1995.

Figures S.2a

Fillers Used in Bombings and Attempted Bombing Incidents, 1976 through 1995

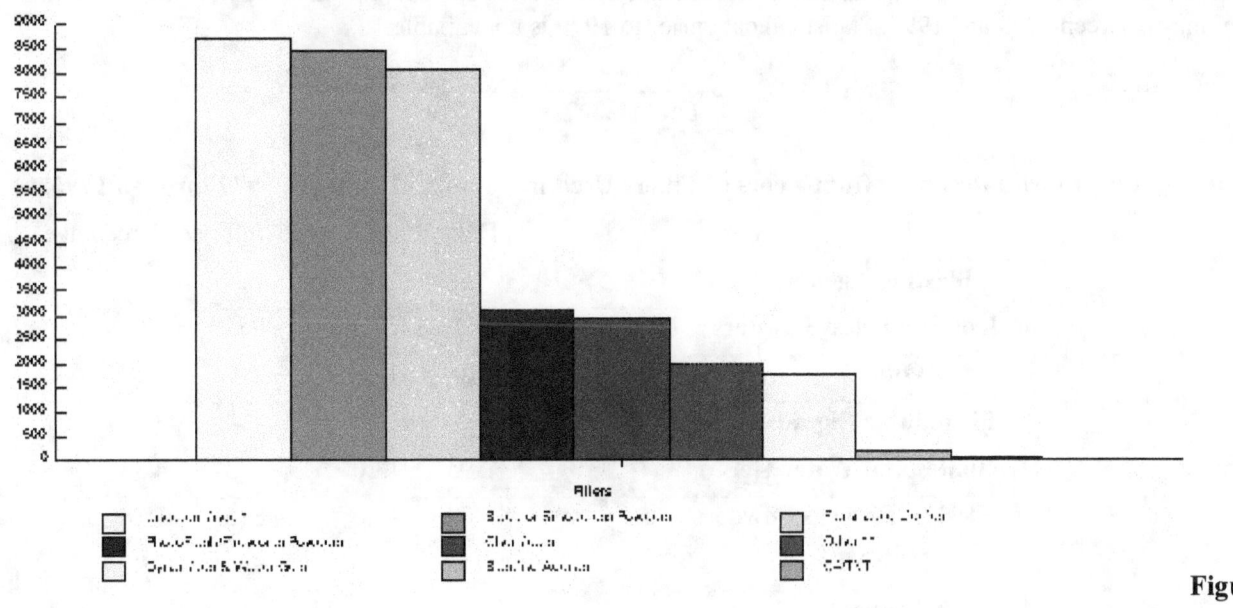

Figure S.2b

Fillers Used in Bombings and Attempted Bombing Incidents, 1991 through 1995

Source:Department of the Treasury, Bureau of Alcohol, Tobacco and Firearms' Explosive Incidents Reports 1985 and 1990, and Arson and Explosive Incidents Reports 1994 and 1995.

·540

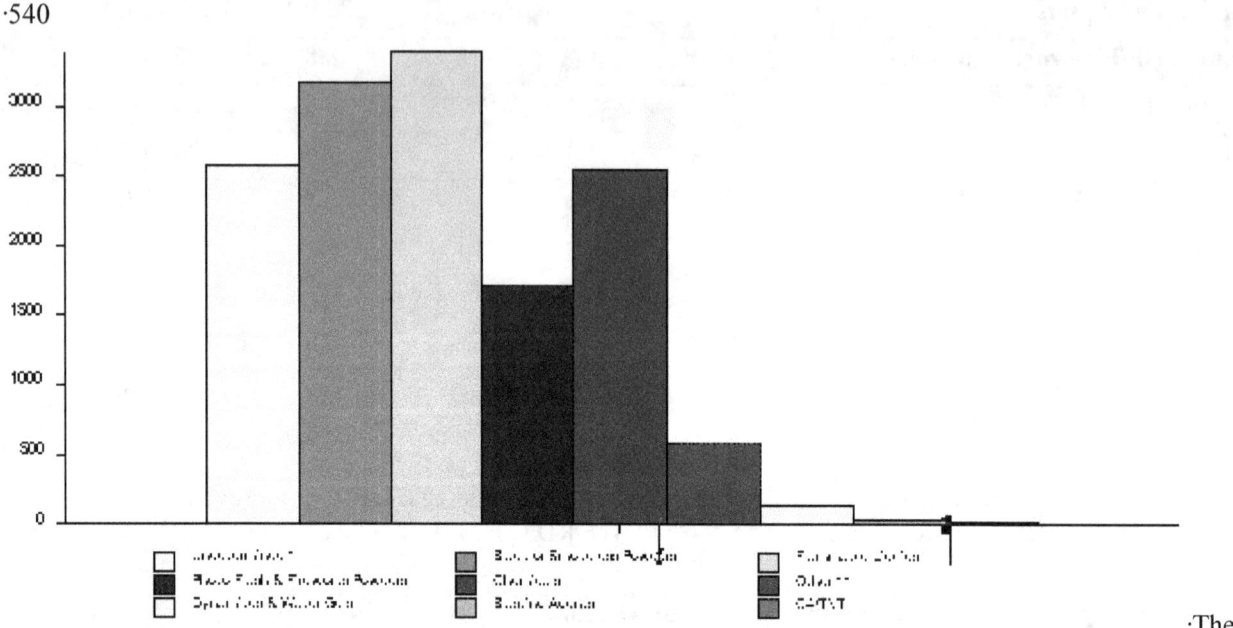

·The Undetermined category captures incidents in which fillers could not be identified through laboratory analysis, or incomplete data was reported.·540

**Other includes match heads, military explosives (excluding C4/TNT), improvised mixtures, flares, boosters, detonating cord, gases, blasting caps, PETN, RDX, HMTD, model rocket propellant, and smoke grenades.

Table S.3, as well as Figures S.3a, and S.3b reflect deaths and injuries attributed to each type of filler used in bombings between 1979 and 1995. Statistical data prior to 1979 is unavailable.

Table S.3				
Deaths and Injuries Resulting from Types of Fillers Used in Bombing Incidents, 1979 through 1995				
Filler-	Deaths	Percentage	Injuries	Percentage
Blasting Agents	175	25.1	536	9.9
Black or Smokeless Powders	140	20.0	957	17.5
Other*	101	14.5	2,417	44.4
Flammable Liquids	99	14.2	474	8.7
Dynamites and Water Gels	95	13.6	233	4.3
Photo Flash/Fireworks Powders	84	12.0	815	15.0
C4/TNT	4	0.6	9	0.2
Chemicals	N/A	N/A	N/A	N/A
Undetermined	N/A	N/A	N/A	N/A
Total	**698**	**100.0**	**5,441**	**100.0**

Source: The Department of the Treasury, Bureau of Alcohol, Tobacco and Firearms, Explosives Incidents

System.

Figure S.3a

Deaths Resulting from Types of Fillers Used in Bombing Incidents, 1979 through 1995

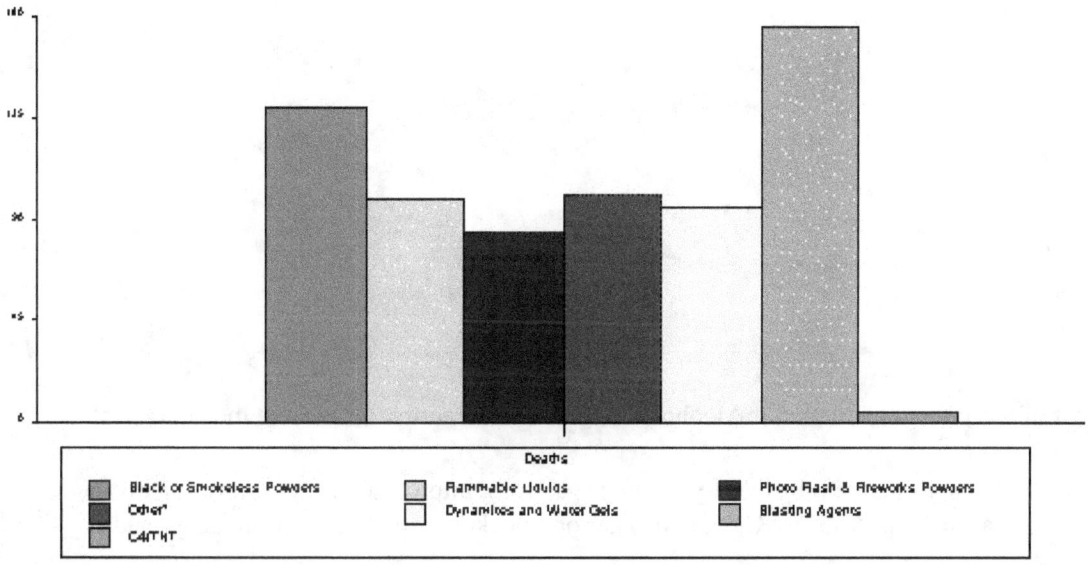

Source: Department of the Treasury, Bureau of Alcohol, Tobacco and Firearms' Explosives Incidents System.

*Other includes match heads, military explosives (excluding C4/TNT), improvised mixtures, flares, boosters, detonating cord, gases, blasting caps, PETN, RDX, HMTD, model rocket propellant, and smoke grenades.

NOTE:For Table S.3, Figure S.3a, and Figure S.3b, as appropriate, the Other category includes six deaths and 1,042 injuries from the World Trade Center Bombing and the Blasting Agents category includes 168 deaths and 519 injuries from the Oklahoma City Bombing.

Figure S.3b

Injuries Resulting from Types of Fillers Used in Bombing Incidents, 1979 through 1995

Undisplayed Graphic

Source: Department of the Treasury, Bureau of Alcohol, Tobacco and Firearms' Explosive Incidents System.

*Other includes match heads, military explosives (excluding C4/TNT), improvised mixtures, flares, boosters, detonating cord, gases, blasting caps, PETN, RDX, HMTD, model rocket propellant, and smoke grenades.

C. CURRENT ATF STUDY

Given the fact that only limited independent testing of taggants has been done in the United States since the OTA completed its report in 1980, the significant issues identified by the 1980 OTA Report had not been resolved as of April 1996, when the Act was enacted. These issues are similar to the ones that the Act mandates the Department to study before deciding whether to propose regulations requiring taggants.

In the interim the Government of Switzerland mandated the use of taggants in explosives used for blasting purposes, other than those mixed on site. In addition, there has been other progress in the research and development of new identification and detection technologies. The Study Group determined that it was necessary to evaluate the progress of past compatibility and safety studies regarding taggants, study the new technologies that have been developed since that date, and examine the experience of other governments in taggant programs.

The efforts of the Study Group are being paralleled and supplemented by independent studies being conducted by the NAS and the International FertilizerDevelopment Center (IFDC). The conclusions and recommendations of the NAS and the IFDC will be assessed in the Study Group's final report. A brief discussion of these independent studies follows.

C-1. Contract with the National Academy of Sciences

ATF entered into a contract with the NAS on August 28, 1996, to conduct an independent and parallel study of the viability of adding tracer elements to explosives for the purposes of detection and identification; the feasibility and practicability of rendering inert common chemicals used to manufacture explosives; and, the feasibility and practicability of imposing controls on certain precursor chemicals used to manufacture explosive materials.

The NAS study will explore and define methods, materials, and technologies that are available today, as well as in research and development, that might be used to enhance the detectability of concealed explosives, or enhance the traceability of illegal explosives after detonation. In making any recommendations for materials to be included as detection or identification elements, the NAS will consider whether the materials pose a risk to human life or safety or substantially impair the quality and reliability of explosives for their intended lawful use. At least three organizations that are capable of conducting testing to validate the study findings will be identified.

The NAS will also evaluate the utility to law enforcement of detection and identification taggants, to include susceptibility to countermeasures, problems of cross-contamination, and ease of detection or identification, analysis and survivability. Materials recommended for inclusion as detection or identification taggants shall not have a substantial adverse effect on the environment. The study shall include an assessment of costs associated with the addition of tracer elements.

The NAS study will explore and define methods, materials, and technologies in use, or in a research and development phase, that might be utilized to render common explosives chemicals inert or less explosive. The NAS study will also explore the feasibility and practicability of imposing controls on certain precursor chemicals used to manufacture explosive materials.

The NAS was directed to provide ATF with a report of its progress, including a description of where its activities will be directed during the remainder of its study, to coincide with this progress report. The NAS has completed its interim report which was released on May 23, 1997. The NAS' final report is to be submitted to ATF no later than 30 days after completion of its study, but not later than February 1998.

C-2. Contract with the International Fertilizer Development Center (IFDC)

On November 5, 1996, ATF entered into a contract with the IFDC to supplement the portion of the study related to the regulation and use of fertilizers. The Act specifically directs that any portion of the study of the regulation and use of fertilizer as a pre-explosive material shall include consultation and input from nonprofit fertilizer centers.

The IFDC was directed to specifically study and report on the feasibility, practicability, and impact of implementing requirements to render common nitrate-based fertilizer chemicals inert, and imposing controls on precursor chemicals used in nitrate-based fertilizers.

The IFDC's report, including its findings and recommendations, will be included in the final report.

C-3. Establishing of ATF-Industry Partnerships

From the onset, the Study Group members recognized the importance of establishing working partnerships with the respective industries and others who have an interest in the results of the Study. To date, the Study Group is working with the following associations, who are among the many with an interest in this study:

The American Pyrotechnic Association

The Chemical Manufacturers Association

The Fertilizer Institute

The International Society of Explosives Engineers

The Institute of Makers of Explosives

The National Rifle Association

The Sporting Arms and Ammunition Manufacturers Institute

As part of ATF's commitment to maintain open lines of communication with these associations and their members, the Study Group has held meetings to discuss the progress of the Study. Additionally, these meetings were intended to provide anopen forum for the associations to present their concerns and share information that may assist ATF in conducting the Study. Letters were also sent to the industry associations requesting any and all information, to include evidence, studies, tests, and other pertinent materials that they possess regarding current and past efforts which may impact the Study. Responses to these letters have been very helpful in conducting the Study.

C-4. Identification of Other Federal and International Efforts

The Study Group has worked to identify, consult with and obtain available information concerning the efforts of other Federal agencies, and international organizations engaged in efforts that have an impact on the issues under study, such as the detection of raw explosives or placing controls on common chemicals for other purposes. Other Federal entities include:

The Department of Defense

The Department of Energy

The Department of Justice

The Department of State's Technical Support Working Group

The Drug Enforcement Administration

The Federal Aviation Administration

The Federal Bureau of Investigation

The General Accounting Office

The International Civil Aviation Organization

The National Institute of Justice

The White House Commission on Aviation Safety and Security

D. TAGGING EXPLOSIVE MATERIALS FOR THE PURPOSES OF DETECTION AND IDENTIFICATION

Pursuant to its mandate to study the tagging of explosive materials, the Study Group visited manufacturers of explosive materials and ammonium nitrate (AN);had limited participation in the testing of taggants; consulted with foreign governments and manufacturers concerning their efforts, including tagging, to address the criminal use of explosives; and received proposals on the testing of existing and new taggant technologies.

D-1. Visits to United States Manufacturers

The Study Group visited several manufacturers of explosive materials. On-site briefings, as well as tours of manufacturing plants, are providing invaluable insight into the manufacturing processes, distribution processes (to include importation and exportation), and legal uses of the materials involved. Consequently, the Study Group will be able to reasonably assess the impact of any changes which may be recommended upon completion of the Study.

·390·On December 11, 1996, the Study Group toured ICI Explosives, Joplin, Missouri, which manufactures explosive-grade AN and emulsions.·390

·390·On December 12, 1996, the Study Group met with representatives, and toured the facilities at Dyno Nobel, Carthage, Missouri. Dyno Nobel is the only domestic manufacturer of nitroglycerin dynamites, and also produces emulsions and slurries.·390

·390·On March 26, 1997, the Study Group toured MP Associates, Ione, California, which designs and manufactures pyrotechnic products/special effects for use by the entertainment industry.·390

D-2. Testing Activities

D-2(a). Dipole Might

Given the fact that questions were raised by the 1980 OTA assessment of the pilot tagging program conducted between 1977 and 1979, the Study Group determined to work with ATF's Dipole Might project to conduct preliminary testing of the survivability and retrievability of developed identification taggants placed in explosive materials. The Dipole Might project, initiated in 1990, provided for the first comprehensive, scientific analysis of large scale (50 - 5,000 pounds) vehicle bombs. It is a multi-national endeavor which uses a computer-aided design program to analyze various effects of vehicle bomb blasts.

However, the completion of the portion of the project that relates to AN has been delayed. A court-approved discovery agreement in the Oklahoma City bombing trial restricted ATF's participation in any tests involving more than 50 pounds ofAN in any improvised explosive devices. Compliance with this court-approved agreement has constrained ATF from participating in tests involving more than 50 pounds of AN to determine the survivability and retrievability of taggants in large AN explosions. As a result, the Dipole Might project is approximately 2 years behind schedule regarding AN testing.

Data generated to date from the Dipole Might project has assisted in the designing of the International Terminal at the San Francisco, California, International Airport, as well as assisting in the evaluation of the vulnerability of the construction of a new U.S. Secret Service building and Department of State embassy facilities. The Dipole Might project is extremely important to ensure that Federal and other public buildings are better able to withstand the effects of terrorist bombings in the future.

D-2(b). Israeli Explosives Tests

From October 14-29, 1996, the Study Group traveled to Israel at the invitation of the Israeli Government to observe and participate in a series of explosives tests, utilizing AN, designed to study blast effects on buildings and automobiles. Due to previously mentioned restrictions on testing AN, the explosive material used by the Israelis was switched from AN to trinitrotoluene (TNT) in both cast and flaked form. The Study Group was able to add taggants to two explosive shots to test the possible survivability of the Microtaggant and Isotag (see Glossary). The substitution however, occurred too late to properly add Microtaggant or Isotag taggants to the TNT to establish an intimate mix and the appropriate ratio of taggants to explosives used.

The primary purpose of the exercise was to conduct Israeli and Dipole Might tests. These tests were conducted; however, the Study Group determined that the taggant portions of the exercise were inconclusive because the taggants were not properly mixed and, therefore, it was impossible to determine if they functioned as designed by the manufacturers.

D-3. Consultations with Foreign Governments and Manufacturers

The Study Group has met with foreign law enforcement agencies and military personnel to discuss their efforts to address the criminal use of explosives and explosive materials, as well as their perspectives on the actual or potential utility of identification markers. The Study Group also met with several foreign manufacturers of explosives and explosive materials. Significant fact-finding activities centered around Switzerland, where the tagging of packaged explosive materials used for blasting purposes has been required by law since 1980.

D-3(a). Switzerland

The Swiss Federal Act of Explosives for Civil Purposes, enacted on July 1, 1980, requires that explosive products specifically designed for blasting be tagged. Bulk explosives which are produced and used on-site (the location where they are to be used) are not required to be tagged.

Representatives of the Swiss Federal Police's Scientific Research Service (SRS), who oversee the Swiss explosives tagging program, met with the Study Group from November 12-15, 1996. During this time, the Study Group also met with the following Swiss manufacturers of explosives, and one Swiss producer of metal powders and alloys (which adds a marker to a premix used in the manufacture of explosives), to discuss their perspectives on the use of markers:

Schweizerische Sprengstoff, Isleten, Switzerland, adds the 3M taggant to dynamite and ammonium nitrate fuel oil (ANFO).

The Societe Suisse des Explosifs, Gamsen, Switzerland, uses the ExploTracer taggant (described below) for water gels, ANFO, dynamite, and black powder, and the 3M taggant for sensitized ANFO and plastic explosives.

Swiss Blasting, Bulach, Switzerland, adds the HF-6 taggant (described below) to slurries and ANFO.

Doral Distribution, Sion, Switzerland, prepares a premix for Swiss Blasting to which the HF-6 taggant is added.

Swiss manufacturers currently use one of three identification taggants; the 3M microchip, HF-6 taggant, and ExploTracer taggant. The 3M (known in the U.S. as the Microtaggant) is a color-coded, polymer microchip consisting of ten layers including a magnetic layer and a fluorescent layer. The HF-6 is similar to the 3M and is coded according to its several layers of color. The HF-6 taggant was developed by Swiss Blasting and is used exclusively in its own products. The ExploTracer taggants consist of synthetic granules (dyed with fluorescent pigments), iron particles, and rare earth elements. The ExploTracer taggant has a limited number of codes and is currently exhausting its capabilities for use in Switzerland.

The Swiss produce approximately 6-11 million pounds of tagged explosives each year. The SRS stated that approximately 40 percent of its annual production of tagged explosives consists of dynamite and the remaining 60 percent is other products such as water gels and slurries.

There are four explosives manufacturers in Switzerland, three manufacturers of high explosives and blasting agents and one manufacturer of low explosives (black powder). The three manufacturers of high explosives produce nitroglycerin-based dynamite, slurries, water gels, plastic explosives, safety fuse, and detonating cord.

Plastic explosives are produced in Switzerland in small quantities. Those produced for the Swiss military contain one of the detection marking agents listed in the 1991 Convention on Marking Plastic Explosives for the Purpose of Detection. Those produced for civil use, primarily surface rock blasting, contain an identification taggant. The Swiss do no produce or import cast boosters for commercial use.

High density (fertilizer grade) AN is not tagged. All three types of identification taggants are used to tag ANFO (manufactured with low density (explosive grade) AN)).

Manufacturers are required to change the taggant code after 6 months or the production of 300 tons of explosive product, whichever occurs first. Production machinery is not cleaned between taggant code changes. The SRS stated that rather than creating a problem of cross-contamination of taggant codes, this absence of cleaning can enhance the ability to trace explosives to a possible purchaser. If two different taggatns are found at an explosion site, the time period during which the explosives were manufactured can be limited to a short period after the first taggant was discontinued and the second taggant was introduced.

Swiss manufacturers estimated that adding taggants to their explosive products has increased the cost of the explosive products by 4-7 cents per pound, depending on the cost of the taggants. They further stated that the administrative costs associated with tagging explosives are minimal; that they did not have to hire additional personnel; and, that they did not have to refit or retool their manufacturing plants.

Swiss manufacturers are required to buy unused explosives back from purchasers (at a lower rate than the original sale price). The returned explosives must be destroyed by open burning or open detonation, or they can be reworked into the production of a new batch of explosives. However, due to concerns over potential cross-contamination in excess of the SRS-established acceptable 10:1 ratio, a new production batch may only contain 6 percent of a returned product. In order to rework explosives, manufacturers must obtain specific authorization from the SRS, and are required to submit samples of the explosives to be reworked to the SRS. Specific data on the percentage of explosives bought back by Swiss

manufacturers is unavailable. However, the SRS estimates the amount to be less than 1 percent, due to the fact that approximately one request by a manufacturer to rework explosives is received per 5-year period.

According to Swiss authorities, controls on the destruction of explosives in Switzerland are not as restrictive as those imposed in the U.S. because the focus is on safety rather than the environmental effects of destruction by detonation or burning. The SRS believes that if burning were the preferred method of destruction, then environmental effects would become an issue. However, as most explosives are destroyed by detonation, this is not the case.

All purchasers of explosives in Switzerland are required to obtain both a working license (issued at the Federal level) and a purchaser's license (issued by the local police). In order to obtain the Federal working license, individuals are required to take a course and pass an examination. In order to apply to take the course, individuals must submit a certificate of reliability, obtained from their local police department, which reflects any criminal convictions or other conditions (such as psychological or substance-abuse problems) which may preclude them from possessing a working license to a Federal commission. This commission, comprised of law enforcement officials and government and commercial explosives experts, approves or denies applications on a case by case basis. Further, if a licensee commits a crime, authorities at the Canton level may revoke his/her license and seize any explosive materials in his/her possession.

There are three types of working licenses, based upon the intended use of the explosives. Purchasing licenses are issued by local law enforcement authorities and are contingent upon inspection of the applicant's storage facilities.

Swiss manufacturers are only liable for their products when they are used for their intended purposes.

The Swiss government, Swiss explosive manufacturers, and importers of explosives into Switzerland reported that there has never been an accidental explosion caused by the addition of taggants to explosives. Further, Swiss manufacturers reported that consumers have not complained that taggants have altered the performance of their explosive products.

Swiss law enforcement officials stated that identification taggants have utility in that they provide them with an additional piece of circumstantial evidence. However, according to Swiss officials, between 1984 and 1994, only 22 percent of all explosives used in bombings were found to be tagged. (Both the Study Group and NAS, with whom ATF has contracted, plan to conduct an in-depth analysis of the correlation between taggants and prosecutions.) After 16 years of using taggants, 78 percent of explosives used in bombings are untagged. The Swiss explained four likely causes for this--

·390·The explosives used were manufactured before the 1980 tagging law went into effect; ·390

·390·Black and smokeless powders manufactured for sporting purposes (which are not required to be tagged) were used; ·390

·390·Military explosives (which are not required to be tagged) were used; or ·390

·390·Explosives from other countries were used. ·390

According to the SRS, the use of pipe bombs containing black and smokeless powders is a significant problem.

D-3(b). Germany

In September 1995, a representative of the German Federal Bureau of Criminal Investigation, Bundeskriminalamt (BKA), made a presentation at the ATF-hosted explosives symposium concerning an electronic, coded detonator system, which the BKA believes may prove to be an effective deterrent to criminal bombings. As a follow-up to this presentation, from November 18-20, 1996, the Study Group met with officials from the BKA who described their current efforts to address the detection and identification of explosive materials, and other strategies for addressing the criminal use of explosive materials.

The current position of the BKA on the marking of explosives for detection and identification is that significant research is required before a viable program can be identified. Further, they consider the detection of explosives to be more important than post-blast identification. The current focus of the BKA in this area is on air traffic safety. Officials stated that they believe detection equipment such as hand-held scanners, currently in use in airports, to be the best available.

Under the sponsorship of the German BKA, the Study Group met with Dynamit Nobel, the manufacturer of an electronic detonator equipped with a microchip that uses a security code to prevent its unauthorized use. The BKA is studying the possible preventative effect of this system, however, the current cost is approximately $20 per detonator. Although the manufacturer indicated that with mass production this cost would decrease, it is currently considered to be cost-prohibitive for the German market. (In the U.S. market electronic detonators, without security microchips, average $2-3.)

D-3(d). Czech Republic - Detection Taggants

On November 22, 1996, the Study Group met officials at the Research Institute for Industrial Chemistry, Pardubice-Sentin, Czech Republic. The Institute is part of Synthesia which manufacturers high and low explosives including Semtex. Semtex, a plastic explosive, and its derivatives have been marked with vapor detectors in accordance with the International Civil Aviation Organization (ICAO) Treaty of 1991. In 1991, Synthesia used ethylene gylcol dinitrate (EGDN) at 0.2 percent as a detection agent. This was discontinued due to ventilation problems on the production line. Synthesia is currently using para-mononitrotoluene (P-MNT) at 0.5 percent or dimethyl-dinitrobutane (DMNB) at 0.1 percent and has not encountered any problems.

D.4 Proposals for Further Study on Identification and Detection Technologies

The Study Group has worked to identify other viable identification and detection technologies. In addition to technical papers presented at the ATF-hosted International Explosives Symposium in September 1995, ATF has received a significant number of proposals for a variety of technologies. The Study Group is currently identifying experts from the scientific community to conduct independent evaluations of these proposals. Further, copies have been provided to the NAS' Committee on Marking, Rendering Inert, and Licensing of Explosive Materials.

Proposals have been received to date from the following developers or vendors of detection and identification technologies. Because many of these proposals are proprietary or pending patents, further descriptive information is withheld from this report.

Biocode, Cambridge, Massachusetts

Centrus Plasma Technologies, Incorporated, Denton, Texas

ChemMech, Riverside, California

Dr. G. Vincent Calder, Racine, Wisconsin

Innovative Biosystems, Moscow, Idaho

Isotag, Houston, Texas

The Los Alamos National Laboratory, Los Alamos, New Mexico

Micro Tracers, Incorporated, San Francisco, California

Micro-Dot Security Systems, Incorporated, Missoula, Montana

Microtrace, Minneapolis, Minnesota

The Oak Ridge National Laboratory, Oak Ridge, Tennessee

The Sandia National Laboratory, Albuquerque, New Mexico

Security Features, McLean, Virginia

SRI International, Menlo Park, California

Systems Support, Incorporated, Great Falls, Virginia

Tracer Detection Technology Corporation, Syosset, New York

TriValley Research, Medford, Oregon

The University of Missouri-Rolla, Rolla, Missouri

The U.S. Army Research Laboratory, Aberdeen Proving Ground, Maryland

West Virginia University Particle Analysis Center, Morgantown, West Virginia

Zia Technologies, New Mexico

D-5. Assessment of Scientific Research and Development Facilities

To date, the following research laboratories have been visited to evaluate their technical expertise and facilities to conduct research, development, and small and large scale testing on those technologies which are assessed as having the greatest potential for successful practical application

The Lawrence Livermore National Laboratory, Livermore, California

The Los Alamos National Laboratory, Los Alamos, New Mexico

New Mexico Tech, Socorro, New Mexico

The Phillips Laboratory, Edwards Air Force Base, California

The Sandia National Laboratory, Albuquerque, New Mexico

The U.S. Army Corps of Engineers, Waterways Experiment Station, Vicksburg, Mississippi

The U.S. Department of the Army, Aberdeen Proving Ground, Aberdeen, Maryland

The U.S. Department of the Navy, Indian Head National Laboratory, Naval Surface Warfare Center, Indian Head, Maryland

E. FEASIBILITY AND PRACTICABILITY OF RENDERING COMMON CHEMICALS USED TO

MANUFACTURE EXPLOSIVE MATERIALS INERT

The Study Group has consulted with chemical industry associations and manufacturers to assess whether common chemicals can be rendered inert.

For purposes of the Study, a common chemical is defined as a widely used chemical compound or element that, as part of an explosive mixture, acts as a fuel or oxidizer in that mixture. Table S.4 contains examples of common chemicals categorized as fuels or oxidizers.

Table S.4			
Chemicals Categorized as Fuels and Oxidizers			
Fuels		**Oxidizers**	
Aluminum Powder	Nitromethane	Ammonium Nitrate	Barium Peroxide
Charcoal	Polyvinylchloride	Sodium Nitrate	Lead Tetroxide
Diesel Fuel	Silicon	Iron Oxide	Potassium Chlorate
Flour	Sugar	Lead Dioxide	Potassium Perchlorate
Iron	Sulfur	Ammonium Perchlorate	Potassium Nitrate
Magnesium	Titanium	Barium Nitrate	Sodium Perchlorate

E-1. Visit to United States Common Chemical Manufacturer

To establish a useful base of knowledge concerning chemical manufacturing and distribution processes, as well as the range of common legal uses for the materials involved, the Study Group conducted research and visited a potassium nitrate manufacturing plant.

·390·On February 24, 1997, the Study Group traveled to Vicksburg Chemical, Vicksburg, Mississippi, which produces potassium nitrate.·390

The research conducted to date demonstrates that common chemicals are fundamental to hundreds of industries. For example, potassium nitrate is marketed as a fertilizer and as tree stump remover, and is also used in the manufacture of products such as optical glass. However, potassium nitrate when combined with fuels such as sucrose (sugar) can also be used as an oxidizer in explosive mixtures.

Elemental sulfur, essential to some explosive products, is used in the manufacture of insecticides, dyes, medicines, and in synthetic and natural rubber. Further, many common organic compounds and powdered metals can readily accept oxygen and act as a fuel in an explosive reaction. Other examples of common chemicals essential to the manufacture of some explosive mixtures can be as diverse as sulfur, charcoal, and aluminum powder.

Of the 24 chemicals listed in Table S.4, only sodium perchlorate and potassium perchlorate have limited non-explosive industry uses. For example, potassium perchlorate is used almost exclusively in explosive pyrotechnic mixtures. All of the others are either used in several different industries or are used in a major capacity in a non-explosive industry. For example, nitromethane, a fuel used in binary explosive products, is used extensively as a model airplane fuel. Ammonium perchlorate, used in some pyrotechnic compositions, is the major oxidizer for solid rocket propellants used for space launches, including satellite launches.

E-2. Ammonium Nitrate as a Common Chemical

Explosive industry representatives report that AN is the most widely used common chemical in the manufacture of commercial explosives. A review of incidents in Great Britain and the U.S. indicates that it has also been used in some of the largest bombings worldwide.

E-2(a). Visits to United States Manufacturers

In addition to contracting with the IFDC for their conduct of an independent study, the Study Group has met with individual manufacturers of AN.

·390·On May 22, 1995, members of the Study Group visited the Mississippi Chemical Corporation, Yazoo City, Mississippi, which manufactures several nitrogen-based fertilizers, including AN.·390

·390·From July 8-10, 1996, the Study Group met with representatives of LaRoche Industries, Cherokee, Alabama, which manufactures AN fertilizer.·390

E-2(b). Consultation with Government of Great Britain

The Study Group met with officials from the British Ministry of Defense to discuss their efforts to desensitize AN. Between 1972-1996, all efforts by the British government to render AN less explosive were countered by certain bomb makers. British officials who have focused on the prevention of AN bombs advise that there is currently no known method which is both feasible and cost-effective by which to desensitize AN and render it non-explosive, while maintaining its effectiveness as a fertilizer.

Based on information currently available to the Study Group, rendering common chemicals, including AN, inert would render them ineffective, in most cases, for their intended purposes. Further, to date it appears that certain additives will make AN fertilizer less sensitive to detonation; however, it can still be detonated with a large booster. An assessment of the British experience, a review of the IFDC study, as well as conversations with explosive and chemical industry representatives and government and private sector scientists indicate that it may be more feasible to establish controls on some common chemicals than to attempt to render them inert.

E-3. Proposals for Further Study on Rendering AN Inert

The Technical Support Working Group (TSWG) has initiated work on assessing various alternatives for desensitizing AN. ATF personnel are included in the TSWG interagency group conducting this assessment. The results of the TSWG study will be available to the ATF Study Group as work progresses.

Finally, outside of the scope of the TSWG and the IFDC studies, the Study Group has received proposals concerning methods to render AN inert without impacting on its effectiveness as a fertilizer. Proposals have been received from the following to date:

The Army Research Laboratory, Aberdeen Proving Ground, Maryland

Arthur D. Little, Cambridge, Massachusetts

The Naval Surface Warfare Center, Indian Head, Maryland

Richard Evons, Boise, Idaho

Sparta, Incorporated, El Segundo, California (including the Phillips Laboratory, the Lawrence Livermore Laboratory, and the University of California at Davis)

The University of Dayton Research Institute, Dayton, Ohio

F. FEASIBILITY AND PRACTICABILITY OF IMPOSING CONTROLS ON CERTAIN PRECURSOR CHEMICALS USED TO MANUFACTURE EXPLOSIVE MATERIALS

For purposes of the Study, a precursor chemical is defined as any chemical compound or element that can be converted to an explosive compound through a chemical reaction or series of reactions; or a chemical compound or element that can catalyze a reaction in which an explosive compound is synthesized. Table S.5 contains examples of precursor chemicals.

Table S.5		
Precursor Chemicals		
Acetone	Iodine	Peroxide
Ammonia	Lead	Silver
Benzene	Mercury	Sulfuric Acid
Butane	Methane	Toluene
Ethylene Glycol	Nitric Acid	Urea
Glycerin	Perchloric Acid	

F-1. Visit to United States Manufacturer

To further its understanding of the chemical industry the Study Group visited a manufacturer of specialty chemicals.

·390·On February 28, 1997, the Study Group met with Olin Chemicals, Brandenburg, Kentucky. This facility is the largest Olin plant dedicated to making specialty chemicals. The product line includes hydraulics and brake fluid, ingredients for cleaning products, and chemical products for theelectronic and micro-electronic industries. ·390

Standard chemical references indicate that no precursor chemicals are without non-explosive applications. Further, research indicates that precursor chemicals are fundamental to hundreds of industries and laboratories. For example:

Nitric and sulfuric acids are used to make nitrated organic high explosives such as nitroglycerin, but they are also used by many other industries, such as in laboratories, in extremely pure form, to test other chemicals.

Acetone is a precursor to a highly sensitive primary explosive, yet it is widely used as nail polish remover.

Many organic (carbon-based) molecules can be nitrated and form a compound with explosive properties.

The Study Group has consulted with the chemical industry and other Federal agencies on ways in which access to some of these chemicals might be better controlled. In particular, it has examined a program in which the chemical industry notifies DEA of substantial purchases of specified chemicals used in the illegal manufacture of narcotics.

G. STATE LICENSING REQUIREMENTS FOR THE PURCHASE AND USE OF COMMERCIAL HIGH EXPLOSIVES

In addition to Federal oversight, manufacturers, dealers, importers, purchasers, and "users" of commercial high explosives are also regulated at the State, county and, in some instances, municipal levels of government, and data has been compiled from these sources. State authorities have supplied the applicable State laws, regulations, codes, statutes, ordinances or legislation, as appropriate. Summary data has also been obtained from the International Society of Explosives Engineers (ISEE), an association representing all facets of the explosives community, and from the Institute of Makers of Explosives (IME), an association primarily representing explosives manufacturers.

State regulation of the legitimate purchase and use of commercial high explosives relates to several broad categories of industry including mining, quarrying, and construction, as well as to those instances wherein commercial high explosives are obtained for non-commercial uses. Some States have separate requirements for commercial and personal application, some have requirements that apply equally to both, and still others have no requirements at all. A number of State officials contacted by representatives of the Study Group advised that there is a possibilitythat their licensing requirements will soon undergo change.

There is some inconsistency in the explosives regulations of the States, the District of Columbia, the Commonwealth of Puerto Rico, and the possessions of the U.S. For example:

Thirty-seven States, the District of Columbia, Puerto Rico, and Guam report that all purchasers must obtain a license or permit to purchase or use commercial high explosives.

Thirteen States report they have no requirement to obtain a license or permit to purchase or use commercial high explosives.

Twenty-four States report the successful completion of training or a test are required to obtain a license or permit to purchase or use commercial high explosives.

H. NEW PREVENTION (DETECTION) TECHNOLOGIES

H-1. Consultation with the Law Enforcement Community

As mandated by law, the Treasury Department contacted the Department of Justice (DOJ) concerning specific new prevention (detection) technologies. DOJ designated an attorney, assigned to its Terrorism and Violent Crime Section, to work with the Study Group. The Study Group has met with the DOJ attorney as well as other representatives from DOJ, the FBI, and the National Institute of Justice (NIJ) to ensure a collaborative effort between the Secretary and the Attorney General to assess detection technologies.

Published evaluations of detection technologies specified in the Act primarily center on studies related to aviation security. Therefore, information gained is generally in the context of the airport environment. Major publications reviewed by the Study Group include the following:

Detection of Explosives For Commercial Aviation Security, prepared in 1993, for the Federal Aviation Administration, by the Committee on Commercial Aviation Security, National Materials Advisory Board,

Commission on Engineering and Technical Systems, National Research Council.

Interim Report to the Federal Aviation Administration Technical Center, prepared in April 1996, as a follow-up to its 1993 report, by the NationalMaterials Advisory Board, Commission on Engineering and Technical Systems, National Research Council.

United States General Accounting Office Report to Congressional Requesters, Terrorism and Drug Trafficking, Technologies for Detecting Explosives and Narcotics, prepared in September 1996.

A project paper prepared by A.N. Garroway and J.B. Miller, of the Naval Research Laboratory, Chemistry Division, entitled Explosives Detection By Pure N NQR which specifically addressed the use of nuclear quadropole resonance to detect explosives.

In addition, the Department of State has established the Technical Support Working Group (TSWG) which has held numerous workshops, of a classified nature, over recent years to identify and report on existing and emerging explosives detection technologies. The TSWG was founded in the 1980s to strengthen U.S. counter-terrorism efforts and improve existing, uncoordinated, unfocused, and inadequate research and development efforts. The TSWG structure was developed to maximize multi-agency input, including ATF's, to identify requirements and prioritize and develop solutions for users. This has become an international effort to provide operationally usable, advanced technology hardware to the counter-terrorism community.

Listed below are the various prevention technologies specifically referred to in the amended Act or otherwise identified by the Study Group.

Vapor/Particle (V/P) detection devices utilize a type of collection system that gathers traces of explosives from a surface or vapors surrounding the target or passenger.

V/P detection devices are non-invasive and able to be used on people. The sample collection step is critical and this technology does not give a direct measure of the amount of explosive present.

The NRC advised the FAA, in its 1996 report, that all detection by trace methods requires further investigation to determine whether or not a threat actually exists. They further advised that a fully functioning system requires integration of sampling, transport, and chemical discrimination. Cost - $60,000- $170,000 per unit depending upon the type of vapor/particle detector utilized.

Computed Tomography (CT) utilizes cross-sectional images integrated by a computer to display density of objects and alarms when characteristics of explosives are detected.

In its 1996 report, the NRC supports CT as the most promising of the x-ray methods. While giving true three-dimensional images, the system's slow speed continues to be a drawback. There are some additional drawbacks, such as the possibility of imaging artifacts, or partial volume effects, that can make it difficult to interpret final data. Cost - $850,000 to $1 million per unit.

Nuclear Quadropole Resonance (NQR) is the examination of nuclei utilizing radio frequency electronics and a computer system. While it is anticipated the equipment could be constructed to detect explosives on a human body, the Technical Support Working Group estimates that this technology could be 2 or more years away.

Thermal Neutron Analysis (TNA) is the capture of low energy neutrons by nitrogen atoms. The resulting de-excitation produces characteristic gamma-rays. It was initially thought to be valuable to detect large quantities of explosives.

TNA is not applicable for use on people because it uses a radioactive source.

NRC disclosed that TNA machines have a high false alarm rate.

Funding for TNA by the FAA was halted in favor of newer technologies that display a greater likelihood of detecting threat quantities of explosives, particularly explosives concealed in baggage.

Pulsed Fast Neutron Analysis (PFNA) utilizes an accelerator to generate neutrons for bombarding a target to measure the induced gamma-rays for the presence of explosives.

PFNA alarms based on three-dimensional images of elemental ratios of hydrogen, oxygen, nitrogen, and carbon.

PFNA determines position and depth of suspect material, but early in its development, the NRC noted that shielding was required, the equipment was large, and it could not be used on people.

NRC reported to the FAA that PFNA is a more likely candidate for successthan TNA but advised that any nuclear detection technology should offer substantial advantages over PFNA due to its high hardware, operational, and maintenance costs. Cost - $8 to $10 million.

Electromagnetic Quadropole Resonance refers to a technology under development in which a radio frequency pulse probes packages, bags, or electrical devices to elicit unique responses from explosives. This system automatically alarms when explosives are detected. This technology is commercially available now, in some forms, at approximately $65,000 per unit.

Nuclear Resonance Absorption (NR-A) refers to a technology under development that may be useful for screening cargo containers. NR-A utilizes an accelerator that generates gamma-rays to penetrate screened objects. The gamma-rays are preferentially absorbed by nitrogen nuclei. A significant decrease in the number of detected gamma-rays indicates the presence of a nitrogen-rich material that could be an explosive. An advantage of this technology is that it requires less shielding than other nuclear technologies. Cost estimates are not available.

Pulsed Fast Neutron Radiography is a technology under development that may be applicable for screening cargo containers. This technology uses an accelerator to generate probing fast neutrons and utilizes the measurement of the transmitted neutron spectrum to detect explosives. Cost estimates are not available.

Remote Hand-Held Millimeter Wave Imaging System, according to information provided to the Study Group at a TSWG workshop in March 1996, is a wide-band, active scanning millimeter wave holographic system that forms high resolution images of concealed objects. These systems utilize illuminating energy that readily penetrates clothing. This developing technology will soon be available for field testing. Cost estimates are not available.

The information available to date indicates that there are many threat scenarios that must be addressed and it is unlikely that one detection technology could be developed that would protect all environments. Further, because there are many prevention technologies to evaluate, and eventually fund and develop, a reliable selection process that encourages wide public participation is critical. The U.S. and foreign countries must continue to share needed information, yet control that information to prevent access by terrorists and other criminal elements. The complexity of the issues presents unique problems that will not easily be solved without further study.

H-2. CANINES

In 1990, ATF began a joint program with the DOS, under its Antiterrorism Assistance Program, to train explosives-detecting canines to be used against terrorism overseas. Under this arrangement, ATF trains the canines for the DOS. These canines are used for different mission needs, which include national security and protection, airport security, etc. To date, ATF has trained over 120 explosives detection canines that are deployed in Israel, Cyprus, Greece, Egypt, Chile, Italy, and Argentina. ATF's Canine Explosives Detection Program trains canines to detect a wider range of explosives in small concentration. The program has been expanded and additional canines are being trained for State and local law enforcement.

With the limitations of the instrument-based technology for explosive detection, trained canines could supplement, and in some instances, more effectively fulfill many of the missions assigned to machines. A major advantage of canines is their mobility, a significant advantage over fixed detector installations which can be circumvented by a terrorist's strategic placement of a bomb.

III. FOCUS OF CONTINUING STUDY

A. TAGGING OF EXPLOSIVE MATERIALS FOR PURPOSES OF IDENTIFICATION AND DETECTION

A-1. Identification Taggants

The utility of identification taggants in helping to produce leads during post-blast investigations of bombings has been clearly demonstrated in the 1979 Baltimore bombing, and potentially through the experience of the Swiss police investigating bombings where taggants were recovered. However, as previously noted, a number of important safety, cost, and environmental issues still need to be resolved.

Reaching definitive conclusions on the appropriateness of including taggants in each type of material that is capable of exploding would be an extremely prolonged task. Although there are literally hundreds of types of explosive materials in the United States, it may not be necessary to review all materials. A review of bombings involving commercial explosives (other than black and smokeless powders) in the United States between 1991 and 1995 reveals that, with few exceptions, they involved only dynamite, slurries, emulsions, or water gels. While dynamite, slurries, emulsions, or water gels are involved in a lesser percentage of overall bombings (2.7 percent or less of incidents which occurred between 1991 and 1995), there have been a number of serious bombings involving those types of explosives in past years. Accordingly, the future work of the Study Group will be focused on these types of explosives. In addition, the Study Group will not examine further the tagging of bulk explosives mixed on-site for immediate use. Law enforcement utility from tagging such explosives is low because there is little risk of theft or other diversion to illegal use. The Study Group will focus on taggant technology currently in use in commercial explosives, and new technologies developed or proposed, such as the Isotag, since the 1980 OTA Report. All currently identified and proposed technologies will be considered, but the evaluation, research, and development of these technologies may continue for an as yet undetermined period of time.

As a further way of focusing its efforts the Study Group, through its research and through its contract with NAS, will use the Swiss experience as a departure point. While the quantities of explosives manufactured are much smaller in Switzerland than the United States and the manufacturing and handling techniques differ, the Swiss program can be viewed as an effective point of reference. The Swiss have been tagging dynamites, slurries, and water gels for 16 years. The Swiss also tag commercially manufactured ANFO and plastic explosives which to date have notbeen a significant problem in U.S. bombings. However, because the Swiss do not conduct underground mining, they neither manufacture nor tag explosive products known as permissibles in the United

States. Permissibles include dynamites, slurries, water gels, and emulsions that have salts added to them to reduce the intensity of flame resulting from the explosion. Very limited compatibility testing has been conducted on permissibles. The Study Group will focus on the differences between the Swiss and U.S. explosives industries, including the absence of permissibles, and attempt to determine whether those differences will adversely affect the safety, utility and cost of taggants in the United States. The Study Group will also examine why, even after 16 years, taggants are recovered in only 22 percent of Swiss bombings and whether there are measures that could be taken in a U.S. taggant program to improve that performance.

A-2. Detection Taggants

Detection taggants could also be of greater utility to law enforcement and public safety. In fact, an even stronger case for their utility might be made since they have the potential to prevent bombings before they occur. Moreover, there is existing technology in wide-spread use for placing detection taggants or markers in plastic explosives. The Study Group will focus its future work on whether it is possible to place the same or similar markers into the most common types of commercial explosives (other than black and smokeless powders) that are involved in U.S. bombings: slurries, emulsions, and water gels. No further study of detection taggants in nitroglycerin-based dynamite will be conducted because it already emits a readily detectable vapor.

B. FEASIBILITY AND PRACTICABILITY OF RENDERING COMMON CHEMICALS USED TO MANUFACTURE EXPLOSIVE MATERIALS INERT

The more than 20 years of experience of the British, the IFDC study, and the Study Group's discussions with the fertilizer manufacturers suggest that there is no current way to render AN inert and still have it function as a cost-effective fertilizer. While the Study Group has received several proposals that purport to render fertilizer inert and is aware of several ongoing research projects in this area, any solutions to this problem are likely to be a number of years in the future. Therefore, as discussed below, the primary focus of the study in the future will be on the utility and feasibility of voluntary or mandatory controls.

The Study Group's review to date of other common chemicals raises similar concerns. These common chemicals are fundamental to hundreds of industries. For example, elemental sulfur is used in the manufacture of insecticides, dyes,medicines, and synthetic and natural rubber. Based upon the Study Group's work to date, there appears to be no known way to render these common chemicals inert and have them retain their intended properties. Accordingly, as with ammonium nitrate, the study will focus on the feasibility and practicability of instituting voluntary or mandatory controls on certain common chemicals.

C. FEASIBILITY AND PRACTICABILITY OF IMPOSING CONTROLS ON CERTAIN PRECURSOR CHEMICALS USED TO MANUFACTURE EXPLOSIVE MATERIALS

Because many chemicals have widespread non-explosive uses, it does not currently appear to be feasible to impose voluntary or mandatory controls on all precursor chemicals. Therefore, the Study Group will focus its efforts on those precursor chemicals that have the greatest utility in the manufacture of explosives.

D. STATE LICENSING REQUIREMENTS FOR THE PURCHASE AND USE OF HIGH EXPLOSIVES

The Study Group has assembled extensive information on State licensing requirements. Review of these requirements shows a wide variety of regulatory schemes in the various States. The focus of the Study Group's future work will be on developing model legislation to provide a uniform approach to requiring a State license or permit and the development of competency standards for "users." In addition, the Study Group will focus on the feasibility of requiring permits for users of all explosives at the Federal level and/or instituting an instant check system.

E. NEW PREVENTION (DETECTION) TECHNOLOGIES

The focus of the Study Group will be to continue to work with the TSWG and identify the most promising detection technologies available.

IV. PLANS AND METHODOLOGY FOR COMPLETING STUDY

The Study Group will take the following actions to achieve the objectives set forth in the focus section above:

A. TAGGING OF EXPLOSIVE MATERIALS FOR PURPOSES OF DETECTION AND IDENTIFICATION

·390·Commission a scientific review to identify the relevant differences in explosives manufacturing and handling techniques between Switzerland and the U. S., determine whether these differences will affect the safety and utility of taggants if they were to be included in dynamite, water gels, emulsions, and slurries in the U.S., and assess vulnerability to countermeasures. The study will also examine the compatibility of taggants with permissibles.·390

·390·Commission an environmental impact assessment to determine the effects of including taggants in dynamite, water gels, emulsions, and slurries, excluding those mixed on-site for immediate use.·390

·390·Commission an independent study to attempt to simulate the July 1979, explosion at the Goex plant in East Camden, Arkansas, and determine whether taggants were the cause of that explosion.·390

·390·Obtain an expert analysis of the costs of including taggants in dynamite, water gels, emulsions, and slurries, excluding those mixed on-site for immediate use.·390

·390·Continue to work with the Swiss Federal Police to further review the use and recovery of taggants in bombings in Switzerland and changes that would be necessary to improve on current results within the United States regulatory environment. For example, the Study Group will continue to explore why taggants have been recovered in 22 percent of all Swiss bombings, and determine if there is a correlation between taggants recovered and prosecutions, and if taggants have had substantial investigative utility for the Swiss.·390

·390·Commission an independent study to determine whether the vapor detection taggants currently used in plastic explosives can be applied to emulsions, water gels, and slurries.·390

B. FEASIBILITY AND PRACTICABILITY OF RENDERING COMMON CHEMICALS USED TO MANUFACTURE EXPLOSIVES MATERIALS INERT

·390·Assemble a panel of experts to review proposals submitted to the Study Group to identify the most promising methods to render AN inert for future research and analysis.·390

·390·Coordinate with the TSWG to obtain the results of its ongoing study into various alternatives for desensitizing AN.·390

C. FEASIBILITY AND PRACTICABILITY OF IMPOSING CONTROLS ON CERTAIN CHEMICALS USED IN THE MANUFACTURE OF EXPLOSIVES MATERIALS

·390·Continue to work with the chemical industry and DEA to determine whether a program like DEA's precursor chemical watch program would be effective to control the diversion of chemicals that are used almost exclusively in the manufacture of explosives.·390

·390·Work with the chemical and explosives industries to identify ways to expand and enhance the voluntary programs for controlling access to materials that can be used to make explosives, similar to the fertilizer industry's "Be Aware for America Program."·390

D. STATE LICENSING REQUIREMENTS FOR THE PURCHASE AND USE OF COMMERCIAL HIGH EXPLOSIVES

·390·Study the feasibility of developing model legislation to provide a uniform approach to State license and permit requirements, develop competency standards, and address the sale of less than case lots of explosives and, if appropriate, draft such legislation.·390

·390·Study the feasibility of requiring a permit for all "users" of all explosives at the Federal level or of instituting an instant criminal history background check system.·390

E. NEW PREVENTION (DETECTION) TECHNOLOGIES

·390·Conduct an international symposium to identify the latest concepts in detection taggant technology that can be used in emulsions, water gels, and slurries.·390

·390·Following the symposium, assemble a group of experts to evaluate the various proposals for detection taggants that have been submitted.·390

·390·Coordinate with TSWG to evaluate those technologies which have the greatest potential for preventing and solving acts of terrorism involving explosive devices.·390

V. SCHEDULE FOR COMPLETING STUDY

The Study Group will issue another report in 1998, and a final report as mandated by the Act within 30 days after conclusion of the Study. At that point, we expect that the action items detailed above will be completed or in stages of significant progress.

GLOSSARY

AMMONIUM NITRATE

Is classified as an oxidizer. An oxidizer is a substance that readily yields oxygen or other oxidizing substances to promote the combustion of organic matter or other fuel. Ammonium nitrate alone is not an explosive material. However, Federal explosives storage regulations require the separation of explosive magazines from nearby stores of ammonium nitrate by certain minimum distances.

ANFO

An explosive material consisting of ammonium nitrate and fuel oil.

BLACK POWDER

A deflagrating or low explosive compound of an intimate mixture of sulfur, charcoal, and an alkali nitrate (usually potassium or sodium nitrate). See LOW EXPLOSIVES.

BLASTING AGENT

Any material or mixture consisting of fuel and oxidizer intended for blasting, not otherwise defined as an explosive, provided that the finished product, as mixed for use or shipment, cannot be detonated by means of a No. 8 test blasting cap when unconfined.

BOOSTER

An explosive charge, usually a high explosive used to initiate a less sensitive explosive. A booster can be either cast, pressed, or extruded.

BULK MIX

A mass of explosive material prepared for use in bulk form without packaging.

COMMERCIAL EXPLOSIVES

Explosives designed, produced, and used for commercial or industrial applications, rather than for military purposes.

COMMON CHEMICALS

Any chemical compound or element that, as part of a physical mixture, would be necessary for that mixture to be considered an explosive mixture; or any chemical compound or element that could be classified as an oxidizer or as a readily available fuel.

C4

A military plastic/moldable high explosive.

DEALER (FEDERAL)

Any person engaged in the business of distributing explosive materials at wholesale or retail.

DETECTION TAGGANTS

A marker or taggant placed into an explosive material that has utility before a bomb explodes.

DETECTION TAGGANTS WITH IDENTIFICATION CAPABILITIES

A marker or taggant placed into an explosive material that has both pre-blast and post-blast utility.

DETONATION

An explosive reaction that moves through an explosive material at a velocity greater than the speed of sound.

DETONATOR

Any device containing an initiating or primary explosive that is used for initiating a detonation. A detonator may not contain more that 10 g of total explosives by weight, excluding ignition or delay charges. The term includes, but is not limited to, electric blasting caps of instantaneous and delay types, blasting caps for use with safety fuses, detonating cord delay connectors, and nonelectric instantaneous and delay blasting caps which use detonating cord, shock tube, or any other replacement for electric leg wires.

DETONATING CORD

A flexible cord containing a center core of high explosive and used to initiate other explosives.

DYNAMITE

A high explosive used for blasting, consisting essentially of a mixture of, but not limited to, nitroglycerin, nitrocellulose, ammonium nitrate, sodium nitrate, and carbonaceous materials.

EMULSIONS

An explosive material containing substantial amounts of oxidizers dissolved in water droplets surrounded by an immiscible fuel.

EXPLOSIVE

Any chemical compound, mixture, or device, the primary or common purpose of which is to function by explosion.

EXPLOSIVES INCIDENTS

This term encompasses actual and attempted explosives/incendiary bombings, stolen, and recovered explosives, hoax devices, and accidental explosions, as defined in ATF's Explosive Incidents Report.

EXPLOSIVE MATERIALS

These include explosives, blasting agents, and detonators. Explosive materials includes, but is not limited to, all items in the List of Explosive Materials. (See Appendix C-1).

EXPLOTRACER TAGGANT

ExploTracer is based on synthetic granules dyed with fluorescent pigments and iron particles. To ensure that each particle has a distinctive code of its own, rare earth elements are added.

FERTILIZER

A substance used to make soil more fertile, such as ammonium nitrate.

FILLER

A type of explosive/incendiary/chemical substance which, in combination with a fusing and/or firing system, constitutes an improvised explosive device (e.g. dynamite, match heads, gasoline).

FLAMMABLE LIQUID

Combustible. A flammable material is one that is ignited easily and burns readily, i.e., gasoline, charcoal lighter fluid, diesel fuel, and paint thinners.

FUEL

Any substance that reacts with the oxygen in the air or with the oxygen yielded by an oxidizer to produce combustion.

HIGH EXPLOSIVES

Explosives which are characterized by a very high rate of reaction, high pressure development, the presence of a detonation wave in the explosive, and which can be caused to detonate by means of a blasting cap when unconfined.

HF-6 TAGGANT

HF-6 is similar to the 3M (Microtaggant) and is coded according to its several layers of color. The HF-6 taggant was developed by Swiss Blasting, and is used exclusively in its own products.

HMTD

An abbreviation for the name of the explosive hexamethylene triperoxide diamine.

IDENTIFICATION TAGGANTS

A marker or taggant placed into an explosive material that has utility after an explosion to identify the manufacturer, the date, and shift when it was manufactured. Once this type taggant is located and identified, the information it provides would allow law enforcement to trace all of the same type explosives manufactured on that specific date and shift to all of the legal purchasers.

IMPORTER

Any person engaged in the business of importing or bringing explosive materials into the United States for purposes of sale or distribution.

INTERSTATE OR FOREIGN COMMERCE

Commerce between any place in a State and any place outside of that State, or within any possession of the United States (not including the Canal Zone) or the District of Columbia, and commerce between places within the same State but through any place outside of that State.

INTRASTATE

Pertaining to or existing within the boundaries of a State of residence.

ISOTAG

A readily identifiable, mass-enhanced, non-radioactive molecular marker that employs the unique chemical structure of the host product without harm to the quality of the product or the environment.

LICENSE (FEDERAL)

Required if a person is intending to engage in the business as an explosive materials manufacturer, importer, or dealer and allows a person to transport, ship, and receive explosive materials in interstate or foreign commerce.

LICENSEE

Any importer, manufacturer, or dealer licensed under the Federal explosives laws.

LOW EXPLOSIVES

Explosives which are characterized by deflagration (a rapid combustion that moves through an explosive material at a velocity less than the speed of sound).

MARKER

See Taggant.

METRIC TON

2,204.6 pounds or 1,000 kilograms.

MICROTAGGANT

Color-coded, polymer microchip consisting of ten layers including a magnetic layer and a fluorescent layer, which is intended to function as an identification taggant. The chip was developed by the 3M Company, but is now manufactured by Microtrace, Minneapolis, Minnesota, which acquired the rights to production in 1984.

NITROGEN (N)

N is one of the three primary plant nutrients, together with phosphorus (P) and potassium (K).

OTHER

For purposes of the EIR, the category of Other includes: match heads, military explosives (excluding C4 and TNT), improvised mixtures, flares, boosters, detonator cord, gases, blasting caps, PETN, RDX, HMTD, model rocket propellant, and smoke grenades.

OXIDIZER OR OXIDIZING MATERIAL

A substance, such as a nitrate, that readily yields oxygen or other oxidizing substances to stimulate the combustion of organic matter or other fuel.

PERMIT

Is required if any person intends to acquire for use explosive materials from a licensee in a State other than the State in which he/she resides, or from a foreign country, or who intends to transport explosive materials in interstate or foreign commerce.

PERMITTEE

Any person who has obtained a Federal User Permit to acquire, ship, or transport explosive materials in interstate or foreign commerce.

PERSON

Any individual, corporation, company, association, firm, partnership, society, or joint stock company.

PETN

An abbreviation for the name of the explosive pentaerythritol tetranitrate.

PHOTOFLASH AND FIREWORKS POWDER

An explosive material intended to produce an audible report and a flash of light when ignited, and typically containing potassium perchlorate, sulfur or antimony sulfide, and aluminum metal.

PRECURSOR CHEMICALS

Any chemical compound or element which can be subjected to a chemical reaction or series of reactions in order to synthesize the chemical compound or element into an explosive compound.

PYROTECHNIC

A chemical mixture which, upon burning, produces visible, brilliant displays, bright lights, or sounds.

RDX

An abbreviation for the name of the explosive cyclonite, hexogen, T4, cyclo-1,3,5,-trimethylene-2,4,6-trinitramine; hexahydro-1,3,5,-trinitro S-triazine.

REWORKED EXPLOSIVES

Any residual or off specification material which can be recycled within the manufacturing process.

SMOKELESS POWDER

Any of a class of explosive propellants that produce comparatively little smoke on explosion and consist mostly of gelatinized cellulose nitrates.

SPECIALTY EXPLOSIVES

Any specialty tool used for a particular purpose other than blasting, such as explosive-actuated device (jet-tappers, jet perforators), propellant-actuated power device (construction nail guns), commercial C-4, detasheet, oil well perforating guns, etc.

SLURRY

An explosive material containing substantial portions of a liquid, oxidizer, and fuel, plus a thickener.

TAGGANT

A solid, liquid, or vapor emitting substance put into an explosive material for the purposes of detection or identification. Also known as a marker or tracer element.

(For purposes of this report, "tagging" is the act of marking or adding a taggant to an explosive material.)

TNT

An abbreviation for the name of the explosive trinitrotoluene.

TON

2,000 pounds or 0.907 metric ton.

TRACER ELEMENT

See Taggant.

UNDETERMINED

For purposes of the EIR, the category of Undetermined captures incidents in which fillers could not be identified through laboratory analysis or incomplete data that was reported.

UREA AMMONIUM NITRATE (UAN)

UAN solution is a popular liquid fertilizer in the United States and other industrialized areas.

USERS

Any persons who purchase and use explosives within their State of residence and are not Federal licensees or permittees.

WATER GEL

An explosive material containing substantial portions of water, oxidizers, and fuel, plus a cross-linking agent which may be a high explosive or blasting agent.

BIBLIOGRAPHY

Atlas Powder Company

Explosives and Rock Blasting ISBN 0-9616284-0-5

Austin Powder Company

Microtaggant Compatibility with Commercial Cast Booster Explosive

Mixture, Report dated February 9, 1994

Bureau of Alcohol, Tobacco and Firearms (ATF)

Arson and Explosives Incidents Reports 1976 through 1995

Bureau of Alcohol, Tobacco and Firearms (ATF)

Compendium of Papers of the International Explosives Symposium,

April 1996,

Bureau of Alcohol, Tobacco and Firearms (ATF)

ATF Form 5400.4 Explosives Transaction Record

Bureau of Alcohol, Tobacco and Firearms (ATF)

ATF P 5400.13 (5/86) Explosives - Federal Agency Directory

Bureau of Alcohol, Tobacco and Firearms (ATF)

Commerce in Explosives: List of Explosive Materials

Bureau of Alcohol, Tobacco and Firearms (ATF)

Federal Explosives Law and Regulations, ATF P 5400.7

Congress Of The United States - Office of Technology Assessment

April 1980 - Taggants in Explosives - Library of Congress

Catalog Number 80-600070

Garroway, A.N. and Miller, J.B.,

Explosives Detection By Pure N NQR on the Use of Nuclear Quadrupolar

Resonance to Detect Explosives - Presented at the First International

Symposium on Explosives Detection Technology Conference,

November 13 - 15, 1991

Institute of Makers of Explosives (IME)

Safety Library Publication Number 12, dated February 1991

Institute of Makers of Explosives (IME)

Compilation of State Laws and Agency Contacts

International Society of Explosives Engineers (ISEE)

Comparison of Explosives Regulations in the Fifty States, August 15, 1994

National Fire Protection Association (NFPA)

NFPA 495 Explosive Materials Code, 1996 Edition

National Research Council

Interim Report to the Federal Aviation Administration

Technical Center, Report Number DOT/FAA/AR-96/51, April 1996

National Research Council

"Detection of Explosives For Commercial Aviation Security," Publication

NMAB-471, National Academy Press, 1993

New Mexico Institute of Mining and Technology

Letter to Austin Powder regarding testing of the Microtaggant, May 9, 1995

Switzerland's Explosives Tagging Program

The New Swiss Act of Explosives and Regulation, July 1, 1980

The Aerospace Corporation Report No. ATR-78(3860-01)-1ND

Explosives Tagging and Control

The American Heritage Dictionary

Second College Edition

The Antiterrorism and Effective Death Penalty Act of 1996 - April 24, 1996

Amended by the Omnibus Consolidated Appropriations Act for Fiscal Year 1996 - September 28, 1996

The International Fertilizer Development Center

Study of Imposing Controls on, or Rendering Inert, Fertilizer Chemicals

Used to Manufacture Explosive Materials - Report - Item B005

March 28, 1997

The Merck Index

Tenth Edition, 1983

U.S. District Court For The Northern District of Texas,

Fort Worth Division - Civil Action Case No. CA 4-80-117E

U.S. Court of Appeals (4th Circuit--Baltimore, Maryland)

U.S. vs Peter McFillin - CR 80-5063

U.S. Department of the Interior (U.S. Geological Survey)

Mineral Industry Surveys - Explosives - 1995

United States General Accounting Office

Report to Congressional Requesters, Terrorism and Drug Trafficking,

Technologies for Detecting Explosives and Narcotics, September, 1996,

GAO/NSIAD/RCED-96-252

Appendix A-1

THE LAW

THE ANTITERRORISM AND EFFECTIVE DEATH PENALTY ACT OF 1996, Approved April 24, 1996,

AS AMENDED BY THE OMNIBUS CONSOLIDATED APPROPRIATIONS ACT OF FISCAL YEAR 1997, Approved on September 30, 1997.

TITLE VII-CRIMINAL LAW MODIFICATIONS TO COUNTER-TERRORISM

SECTION 732, MARKING, RENDERING INERT, AND LICENSING OF EXPLOSIVE MATERIALS.

(a) STUDY. -

(1) IN GENERAL. -- Not later than 12 months after the date of enactment of this Act, the Secretary of the Treasury (referred to in this section as the "Secretary") shall conduct a study of --

(A) the tagging of explosive materials for purposes of detection and identification;

(B) the feasibility and practicability of rendering common chemicals used to manufacture explosives materials inert;

(C) the feasibility and practicability of imposing controls on certain precursor chemicals used to manufacture explosive materials; and

(D) State licensing requirements for the purchase and use of commercial high explosives, including --

(i) detonators;

(ii) detonating cords;

(iii) dynamite;

(iv) water gel;

(v) emulsion;

(vi) blasting agents; and

(vii) boosters

(2) EXCLUSION. -- No study conducted under this subsection or regulation proposed under subsection (a) shall include black or smokeless powder among the explosive materials considered.

(3) New prevention technologies: In addition to the study of taggants as provided herein, the Secretary, in consultation with the Attorney General, shall concurrently report to the Congress on the possible use, and exploitation of technologies such as vapor detection devices, computed tomography, nuclear quadropole resonance, thermal neutron analysis, pulsed fast-neutron analysis, and other technologies upon which recommendations to the Congress may be made for further study, funding, and use of the same in preventing and solving acts of terrorism involving explosive devices.

(b) CONSULTATION.

(1) IN GENERAL. -- In conducting the study under subsection (a), the Secretary shall consult with --

(A) Federal, State, and local officials with expertise in the area of chemicals used to manufacture explosive materials; and

(B) such other individuals as the Secretary determines are necessary.

(2) FERTILIZER RESEARCH CENTERS.-- In conducting any portion of the study under subsection (a) relating to the regulation and use of fertilizer as a pre-explosive material, the Secretary of the Treasury shall consult with and receive input from non-profit fertilizer research centers.

(c) REPORT.-- Not later than 30 days after the completion of the study conducted under subsection (a), the Secretary shall submit a report to the Congress, which shall be made public, that contains --

(1) the results of the study;

(2) any recommendations for legislation; and

(3) any opinions and findings of the fertilizer research centers.

(d) HEARINGS.--Congress shall have not less than 90 days after the submission of the report under subsection (c) to —

(1) review the results of the study; and

(2) hold hearings and receive testimony regarding the recommendations of the Secretary.

(e) REGULATIONS.--

(1) IN GENERAL .-- Not later than 6 months after the submission of the report required by subsection (c), the Secretary may submit to Congress and publish in the Federal Register draft regulations for the addition of tracer elements to explosive materials manufactured in or imported into the United States, of such character and in such quantity as the Secretary may authorize or require, if the results of the study conducted under subsection (a) indicate that the tracer elements --

(A) will not pose a risk to human life or safety;

(B) will substantially assist law enforcement officers in their investigative efforts;

(C) will not substantially impair the quality of the explosive materials for their intended lawful use;

(D) will not have a substantially adverse effect on the environment; and

(E) the costs associated with the addition of the tracers will not outweigh benefits of their inclusion.

(2) EFFECTIVE DATE.-- The regulations under paragraph (1) shall take effect 270 days after the Secretary submits proposed regulations to Congress pursuant to paragraph (1), except to the extent that the effective date is revised or the regulation is otherwise modified or disapproved by an Act of Congress.

(f) SPECIAL STUDY:

(1) In general.--Notwithstanding subsection (a), the Secretary of theTreasury shall enter into a contract with the National Academy of Sciences (referred to in this section as the "Academy") to conduct a study of the tagging of smokeless and black powder by any viable technology for purposes of detection and identification. The study shall be conducted by an independent panel of 5 experts appointed by the Academy.

(2) Study elements.--The study conducted under this subsection shall--

(A) indicate whether the tracer elements, when added to smokeless and black powder--

(i) will pose a risk to human life or safety;

(ii) will substantially assist law enforcement officers in their investigative efforts;

(iii) will impair the quality and performance of the powders (which shall include a broad and comprehensive sampling of all available powders) for their intended lawful use, including, but not limited to the sporting, defense, and hand loading uses of the powders, as well as their use in display and lawful consumer pyrotechnics;

(iv) will have a substantially adverse effect on the environment;

(v) will incur costs which outweigh the benefits of their inclusion, including an evaluation of the probable production and regulatory cost of compliance to the industry, and the costs and effects on consumers, including the effect on the demand for ammunition; and

(vi) can be evaded, and with what degree of difficulty, by terrorists or terrorist organizations, including evading tracer elements by the use of precursor chemicals to make black or other powders; and

(B) provide for consultation on the study with Federal, State, and local officials, non-governmental organizations, including all national police organizations, national sporting organizations, and national industry associations with expertise in this area and such other individuals as shall be deemed necessary.

(3) Report and costs.--The study conducted under this subsection shall be presented to Congress 12 months after the enactment of this subsection and be made available to the public, including any data tapes or data used to form such recommendations. There are authorized to be appropriated such sums as may be necessary to carry out the study.
   This was last updated on August 25, 1998

www.ingramcontent.com/pod-product-compliance
Lightning Source LLC
Chambersburg PA
CBHW081625170526
45166CB00009B/3104